RAND NATIONAL SECURITY RESEARCH DIVISION

Current and Future Exposure of Infrastructure in the United States to Natural Hazards

Henry H. Willis, Anu Narayanan, Jordan R. Fischbach, Edmundo Molina-Perez, Chuck Stelzner, Kathleen Loa, Lauren Kendrick

Prepared for the U.S. Department of Homeland Security, Office of Infrastructure Protection

Approved for public release; distribution unlimited

For more information on this publication, visit www.rand.org/t/rr1453

Library of Congress Cataloging-in-Publication Data
ISBN: 978-0-8330-9500-8

Published by the RAND Corporation, Santa Monica, Calif.
© Copyright 2016 RAND Corporation
RAND® is a registered trademark

Cover Image: © Fotolia/kentoh

Support RAND
Make a tax-deductible charitable contribution at
www.rand.org/giving/contribute

www.rand.org

Preface

The Department of Homeland Security, National Preparedness and Programs Directorate, Office of Infrastructure Protection, has asked the RAND Corporation to analyze exposures of national infrastructure systems to natural hazards and how these exposures are expected to evolve in response to climate change. This report is one of two that document this work. This report describes insights about exposures from natural hazards now and in the future, as well as gaps in data that, if filled, could improve the nation's ability to assess infrastructure risk and improve infrastructure resilience. The data, models, and methods used in this analysis are described in detail in:

> Anu Narayanan, Henry H. Willis, Jordan Fischbach, Drake Warren, Edmundo Molina-Perez, Chuck Stelzner, Katie Loa, Lauren Kendrick, Paul Sorenson, and Tom LaTourrette, *Characterizing National Exposures to Infrastructure from Natural Disasters: Data and Methods Documentation*, Santa Monica, Calif.: RAND Corporation, RR-1453-1-DHS, 2016.

These reports should be of interest to scientists, planners, and policymakers interested in climate change adaptation and community resilience.

These research efforts were conducted in the Homeland Security and Defense Center (HSDC), which conducts analysis to prepare and protect communities and critical infrastructure from natural disasters and terrorism. Center projects examine a wide range of risk-management problems, including coastal and border security, emergency preparedness and response, defense support to civil authorities, transportation security, domestic intelligence, and technology acquisition. Center clients include the U.S. Department of Homeland Security, the U.S. Department of Defense, the U.S. Department of Justice, and other organizations charged with security and disaster preparedness, response, and recovery.

HSDC is a joint center of two research divisions: RAND Justice, Infrastructure, and Environment and the RAND National Security Research Division. RAND Justice, Infrastructure, and Environment is dedicated to improving policymaking and decisionmaking in a wide range of policy domains, including civil and criminal justice, infrastructure protection and homeland security, transportation and energy policy, and environmental and natural resource policy. The RAND National Security Research Division conducts research and analysis for all national security sponsors other than the U.S. Air Force and the Army. The division includes the National Defense Research Institute, a federally funded research and development center whose sponsors include the Office of the Secretary of Defense, the Joint Staff, the Unified Combatant Commands, the defense agencies, and the U.S. Department of the Navy. The National Security Research Division also conducts research for the U.S. Intelligence Community and the ministries of defense of U.S. allies and partners.

Questions or comments about this report should be sent to the project leader, Henry Willis (hwillis@rand.org). For more information about the Homeland Security and Defense Center, see www.rand.org/hsdc or contact the director at hsdc@rand.org.

Contents

Figures and Tables

Figures

Tables

Summary

Communities, companies, and governments at all levels are making decisions that will influence where, what, and how infrastructure will be built. These design and policy decisions shape infrastructure, influence economic development, and influence future exposures to natural hazards for decades. If communities are to manage infrastructure resilience, these decisions must be supported by analysis and stakeholder processes that reflect the best knowledge of current and future natural hazard exposures, infrastructure vulnerabilities, and the consequences of disruptions—i.e., infrastructure risks.

Population growth and shifts, particularly those to the coasts, drive demand for new infrastructure and, as a result, increase the exposure of infrastructure to natural hazards. These natural hazard exposures are projected to be larger and more uncertain in the future because of the effects of sea level rise and projected changes in temperature and precipitation patterns. Thus, incorporating natural hazard risk assessment into infrastructure planning is becoming both increasingly important and challenging.

In some cases, infrastructure planning is supported by intensive analysis and stakeholder processes. Typically, this includes infrastructure projects that involve billions of dollars. Often, the intensive processes are supported by external resources or court mandates. Absent such disaster recovery funds or special compensation mechanisms, small and medium-sized communities do not typically have the resources to conduct risk assessments to understand future natural hazard exposures and develop tools for analysis and processes to support local infrastructure planning. In these cases, communities need data about where infrastructure exists, what the intensity and likelihood of natural hazards are today, and how natural hazards will change in the future.

RAND's Analysis of Infrastructure Exposure

The Department of Homeland Security, Office of Infrastructure Protection, is responsible for assessing the state of the nation's critical infrastructure protection and resilience; supporting the development of national infrastructure protection strategy and policy; and leading organizational strategic planning, performance measurement, and budgeting activities. To support this work, the Office of Infrastructure Protection asked RAND to analyze infrastructure exposure to natural hazards in the continental United States, and to the extent possible, analyze how climate change may impact future exposure.

In response, we collected data about infrastructure exposures to natural hazards across the continental United States, developed projections for how these exposures are expected to change in the future as a result of climate change, and used these data to analyze infrastructure exposures to natural hazards in the United States. Unlike many prior assessments, which have focused on individual infrastructure sectors or on individual hazards, the analysis described in this report and the accompanying documentation on underlying data and methods provide an integrated view of infrastructure exposure to a range of potentially high-intensity natural hazards. The analysis in this report describes exposures across multiple infrastructure sectors and multiple hazards, and it incorporates both current infrastructure exposure to natural hazards and uncertainty about the extent and intensity of future natural hazards in the context of projected changes to climate patterns.

Findings and Implications for Policy and Research

Several key findings emerge from this analysis:

- Infrastructure exposure to natural hazards is expected to increase—in some cases substantially—across the continental United States.
- Infrastructure in some areas of the country already faces disproportionate exposure to natural hazards, and this exposure is likely to increase in the future as a result of climate change.

These findings are likely to suggest several implications for policy and future research. In the short term, several are apparent:

- Infrastructure and community resilience efforts should incorporate potential impacts from climate change and resulting increases in exposure to natural hazards.
- More granular information is needed about specific natural hazard exposure and the specific infrastructure communities will need to respond effectively to climate change-induced natural hazard exposure changes.

The exposures presented in this report are lower bound on current and future exposures due to limits in infrastructure data, natural hazard data, and understanding of the effects of climate change. The most significant data gaps include

- incomplete baseline data for several hazards and areas, in particular for riverine flooding and areas where coastal flood maps are not available, which include the state of Louisiana and parts of Florida
- lack of probabilistic exposure data for several hazards, especially ice storms, drought, and wildfires.
- uncertainty about the effects of climate change on many hazards, especially riverine flooding and hurricane winds
- inadequate detail for infrastructure data to support risk analysis
- lack of complete data for Alaska, Hawaii, and U.S. territories
- uncertainty about where and how future infrastructure development will occur.

Each of these observations provides a basis for planning as the federal government, state and local governments, and the private sector seek to improve resilience against current and future exposure to high-intensity natural hazards. Addressing these issues effectively and comprehensively will require multifaceted efforts that include collecting descriptive information from and about communities, improving scientific knowledge about hazard phenomena, and developing tools and institutions to plan mitigation strategies for the complex and uncertain array of natural hazards that do and may increasingly threaten communities across the nation.

Acknowledgments

This work has benefited from the thoughtful reviews and suggestions from several individuals including Lisa Barr (Department of Homeland Security), Cynthia Cook (RAND), Paul DeLuca (RAND), Sarah Ellis Peed (Department of Homeland Security), David Groves (RAND), Seth Guikema (University of Michigan), Debra Knopman (RAND), Robert Kolasky (Department of Homeland Security), Merideth Secor (Department of Homeland Security), and Kate White (U.S. Army Corps of Engineers).

CHAPTER ONE

Introduction: The Need to Better Understand Current and Future Hazard Exposure

Following 2005's Hurricane Katrina, the state of Louisiana developed a master plan to guide $50 billion in infrastructure investment in coastal reconstruction (LACPRA, 2012). Following 2012's Superstorm Sandy, the federal government provided almost $12 billion to restore critical infrastructure. These recent experiences are a small fraction of the burden involved with ensuring that our nation has the infrastructure required to support economic security.

The American Society of Civil Engineers estimates that $3.6 trillion of investment in U.S. infrastructure is required through 2020 (ASCE, 2013). These investments are needed to address multiple pressures on critical infrastructure. Use and time increase wear on infrastructure systems, many of which were built throughout the 20th century. At the same time, the emergence of new technologies creates the pressure and opportunity to modernize infrastructure and increase its efficiency and productivity. Economic development, population growth, and population migration within the United States are increasing and shifting demand for infrastructure services.

Governments at all levels are responding to these infrastructure needs and making decisions that will influence where, what, and how infrastructure will be built. Presidential Policy Directive 21 on Critical Infrastructure Security and Resilience (White House, 2013) and the subsequent 2013 National Infrastructure Protection Plan (DHS, 2013) outline roles and relationships through which the government and private sector invest, plan, and act to improve the security and resilience of infrastructure systems.

Actions by other federal departments reinforce this policy guidance across agencies. In the context of planning, the Department of Energy's first Quadrennial Energy Review identifies the current and future threats, risks, and opportunities for U.S. energy transmission, storage, and distribution infrastructure (DOE, 2015), with special attention being paid to defining and measuring resilience (Willis and Loa, 2015). In the context of economic development, the Department of Housing and Urban Development's Hurricane Sandy Rebuilding Task Force established guidelines for federal investment in infrastructure to better protect communities and improve community resilience (HUD, 2013; Finucane et al., 2104). Communities are weighing current and future threats as part of infrastructure planning and design (LACPRA, 2012; NYCEP 2014).

These design and policy decisions will shape infrastructure, influence economic development, and influence future exposures to natural hazards for decades. If communities are to manage infrastructure resilience, these decisions must be supported by analysis and stakeholder processes that reflect the best knowledge of natural hazard exposures, infrastructure vulnerabilities, and the consequences of disruptions; in other words, infrastructure risks.

Population growth and shifts, particularly those in the Atlantic, Gulf, and Pacific coasts, increase the exposure of infrastructure to natural hazards. These natural hazard exposures are projected to be larger and more uncertain in the future because of the effects from rising sea levels and projected changes in temperature and precipitation patterns. Thus, incorporating natural hazard risk assessment into infrastructure planning is becoming both increasingly important and challenging.

In some cases, infrastructure planning is supported by intensive analysis and stakeholder processes. Typically this entails infrastructure projects that involve billions of dollars. More often, the intensive processes are supported by external resources or mandated by courts. Examples include the planning following Hurricane Katrina, supported in part by compensation funds from the Deepwater Horizon oil spill; reconstruction in New York and New Jersey, supported by the Sandy recovery funding; or long-range water planning in Southern California and the Colorado River Basin (Groves, et al. 2008; Groves et al., 2013).

However, the need for risk-informed infrastructure planning extends beyond these special situations. Absent disaster recovery funds or special compensation mechanisms, small- and medium-sized communities do not typically have resources to conduct risk assessments to understand future natural hazard exposures and develop analysis tools and processes to support local infrastructure planning. These communities need data about where infrastructure exists, what the intensity and likelihood of natural hazards are today, and how these exposures will change in the future.

RAND's Analysis of Infrastructure Exposure

The Department of Homeland Security, Office of Infrastructure Protection, Office of Strategy, Policy and Budget, is responsible for assessing the state of America's critical infrastructure protection and resilience programs; supporting the development of national infrastructure protection strategy and policy; and leading organizational strategic planning, performance measurement, and budgeting activities. To support this work, the Office of Infrastructure Protection asked RAND to analyze infrastructure exposure in the continental United States to natural hazards, and to the extent possible, analyze how climate change may impact future exposure.

In response, we collected data about infrastructure exposures to natural hazards across the continental United States, developed projections for how these exposures are expected to change in the future as a result of climate change, and used these data to analyze natural hazard exposures to infrastructure in the United States. The data sets and analysis methods applied are summarized in the next section. In addition, the data sets and technical aspects of the analysis are fully documented in a separate technical report and available to federal, state, and local governments and U.S. Department of Homeland Security (DHS) infrastructure partners through the DHS Office of Infrastructure Protection (Narayanan et al., 2016).

This report summarizes insights we have gained about the exposures to U.S. infrastructure from natural hazards now and in the future. Our analysis identifies regions in the country where infrastructure may be uniquely exposed to a complex set of natural hazards. In those regions, our analysis highlights the types of infrastructure that are exposed and the hazards that put them at risk. Our analysis also reveals where infrastructure exposures may be expected to change most dramatically. Finally, our analysis reveals where infrastructure exposures remain most uncertain and where new data and analysis would be most valuable. Each of these findings can inform federal efforts to improve infrastructure and resilience planning.

CHAPTER TWO

Defining and Analyzing Infrastructure Exposure

In order to analyze infrastructure exposure, it is important to first define what it is and define how infrastructure exposure relates to infrastructure risk. These definitions guide what and how data are used to characterize exposure.

Defining Exposure

The National Infrastructure Protection Plan defines *risk* as a function of threat, vulnerability and consequence (DHS, 2013). *Threats* from natural hazards depend on whether the conditions that produce the hazard are present at a location. Because of the uncertainty in natural hazard events, such threats are described with measures of both intensity and likelihood.[1] For example, hurricane winds, storm-driven flooding, and earthquakes are described, respectively, as sustained wind speeds with a 100-year return interval, flood depths with a 500-year return interval, and peak ground acceleration from seismic activity with a 2,500-year return interval. *Risk* exists if exposure to a natural hazard exists. Accordingly, for purposes of this report, infrastructure is *exposed* to a natural hazard when the likelihood of a natural hazard occurring at an intensity that could cause significant damage or disruption exceeds a specific threshold.

Vulnerability is a state of an infrastructure asset or system that is dependent on how the infrastructure is designed and operated and how these two factors affect its response to exposure to a hazard or hazards (Haimes, 2006). Thus, assessing the vulnerability of a given piece of infrastructure requires detailed knowledge of the infrastructure's current condition, as well as a fundamental understanding of the interaction between the infrastructure and specific hazards at different levels of intensity. For example, the vulnerability of a power plant to hazards will depend on whether wind, flood, or seismic mitigation steps have been incorporated into the facility design; whether regular maintenance schedules have been followed; and whether maintenance resulted in upgrades of the facility to current hazard mitigation standards.

Consequences of infrastructure exposure result from physical damage to the infrastructure, disruptions of the services that infrastructure systems provide, and how communities may need to respond to and recover from these service disruptions (Willis and Loa, 2015).

---

[1] Uncertainty in the occurrence of natural hazards is commonly expressed in terms of event return periods, i.e., the average time between occurrences of events. A 100-year return period is an event that occurs with an average interval of 100 years between events. It is equivalent to an event with a 1 in 100 (1 percent) chance of occurring within a given year. It does not mean that the event would only occur once every 100 years, or that the event would occur at all within a 100-year period.

Assessing consequences requires detailed knowledge of the nature and severity of disruptions, as well as the economic and physical effects of disruptions on communities.

Infrastructure data currently available to support nationwide analysis contain significant gaps. These data gaps, which are described later in this report, limit the ability to describe nationwide infrastructure vulnerabilities to and consequences from natural hazards. As a result, this analysis does not address risk and is instead limited to exposure to natural hazards. The data and methods we used in this analysis are fully documented by Narayanan et al. (2016) and summarized in this chapter.

We define exposure as existing when a hazard occurs in the same location where an infrastructure asset is present, and the infrastructure asset could be impacted by the hazard (i.e., the infrastructure is potentially vulnerable to the hazard). However, the analysis accounts for vulnerability only to the extent that there is the potential for a direct physical interaction between the type of infrastructure and the type of hazard (e.g., underground pipelines are not exposed to wind because of their innate physical characteristics, regardless of their current state or condition). The interactions between infrastructure types and hazards are listed in Appendix.

Data and Analytical Approach

Infrastructure data used in this analysis are drawn from the 2013 Homeland Infrastructure Foundation-Level Data Homeland Security Infrastructure Protection (HSIP) Gold data set.[2] We analyzed a subset of infrastructure contained in this database that would lead to the most significant consequences if it were affected by a natural hazard. Our screen for consequences focused on those infrastructure sectors that, if affected, would either result in major direct property or casualty consequences as a result of damage to a specific facility, extend within a sector regionally, or cascade to effects in other sectors. Because this data set contains proprietary commercial information, this analysis describes infrastructure exposure to natural hazards aggregated to the county level so as to not reveal any of the underlying proprietary information. Table 2.1 lists the sectors and subsectors we selected using these criteria.

Exposure depends on the likelihood and intensity of hazards in question. So, to allow for analysis of exposure across a range of hazard intensities and likelihoods, we collected data sets for 11 natural hazards that could affect infrastructure at a regional scale (see Table 2.2). Ideally we sought data sets that provided nationwide coverage, provided a measure of hazard intensity, and characterized changes in likelihood at different intensity levels. For example, the data set for hurricane winds describes the peak sustained wind speeds at return periods of 10, 20, 50, 100, 200, 500, and 1,000 years. Return periods provided for each hazard are listed in Table 2.3.

The effects of climate change are expected to modify the intensity, frequency, and geographic patterns of some natural hazards. For example, sea level rise exacerbates coastal flooding. Changes in precipitation modify riverine flooding. Changes in temperature and shifts in precipitation patterns can change the nature of ice storms, drought, tornadoes, and wildfires. Changes in ocean temperatures can affect hurricane strength and frequency.

---

[2]   For more information, see the Homeland Infrastructure Foundation–Level Data Subcommittee Online Community web page (undated).

Table 2.1
Number of Total Assets in Each Infrastructure Layer

Infrastructure Sector	Subsectors	No. Included in Analysis
Chemical industry	Chemical manufacturing	52,759
Communications	Internet exchange points	78
Energy (including nuclear power)	Electric power generation plants	4,017
	Electric power substations	870
	Energy power transmission lines[a]	208,612
	Energy distribution and control facilities	80
	Natural gas imports/exports points	26
	Natural gas processing plants	179
	Nuclear fuel facilities	13
	Nuclear plants	89
	Oil and natural gas pipelines[a]	1,685,806
	Petroleum, oil, and lubricants storage facilities	58
	Refineries	144

Table 2.1—Continued

Infrastructure Sector	Subsectors	No. Included in Analysis
Transportation	Airports[a]	180
	Canals[a]	1,186
	DHS-identified railroad bridges	114
	Railroad stations	499
	Railroad transit lines[a]	73,528
	DHS-identified railroad tunnels	30
	Railroad yards	2,211
	DHS-identified road bridges and tunnels	140
	Coastal, Great Lakes, and inland ports	22,635
	FAA air route traffic control centers	22
	Fixed-guideway transit systems, stations, and lines[a]	5,340
	Intermodal terminal facilities	3,270
	Interstate highways[a]	83,443
	Locks	219
Water supply and wastewater treatment	Dams	372
	Wastewater treatment plants	3,970

[a] = Indicates estimated number of infrastructure points. Point counts were estimated when the original geospatial information systems (GIS) infrastructure data set type were in line or area form. In this case, centroid or distance rules were applied for point creation. FAA = Federal Aviation Administration.

Table 2.2
Hazards Included in RAND Analysis

Type of Hazard	Hazard	Measure of Intensity
Climate-adjusted hazards	Coastal flooding (including permanent inundation, tidal flooding, and storm surge)	Flood depth
	Extreme temperature	Monthly maximum temperature
	Meteorological drought (dryness)	Keetch-Byram Drought Index (KBDI) value
	Wildfires	Wildfire potential index value
Climate-unadjusted hazards	Earthquakes	Peak ground acceleration
	Hurricane winds[a]	Peak sustained winds
	Ice storms[a]	Sperry-Piltz Ice Accumulation Index value
	Landslides	U.S. Geological Survey landslide susceptibility assessment
	Riverine flooding[a]	Location of 100-year flood plain
	Tsunamis[a]	Coastal elevation and categorization by the Risk Management Solutions tsunami hazard assessment
	Tornadoes[a]	Peak sustained winds

[a] = indicates hazards for which climate change could affect the intensity or likelihood of exposure across the United States, but data and models are not available to characterize these changes.

Table 2.3
Return Periods Associated with Each Hazard

Hazard	Return Periods
Coastal flooding	2/5/10/20/50/100 years
Extreme temperature	2/5/10/20/50/100 years
Meteorological drought (dryness)	75th and 95th KBDI values
Wildfires	Not applicable
Earthquakes	500 and 2,500 years
Hurricane winds	10/20/50/100/200/500/1,000 years
Ice storms	50 years
Riverine flooding	100 years
Tsunamis	≤ 500 years
Tornadoes	100,000 years
Landslides	Not applicable

While potential links between climate change and natural hazards are widely recognized, data and models that project changes in natural hazard exposures vary across hazards. In some cases, models and data exist to support adjustments of baseline hazard data to reflect climate change. These climate-adjusted hazards, listed in Table 2.2, include coastal flooding, extreme temperature, drought, and wildfire. The approach used to represent climate scenarios for these hazards is described in more detail in Chapter Four.

For other hazards, climate change is expected to modify exposure, but data and models do not exist to explain how exposure will change. For example, changes in climate could affect riverine flooding, hurricane winds, ice storms, tornadoes and tsunami exposures. At the same time, some hazards, such as earthquakes and landslides, are not likely to be affected by climate change. For all of these hazards, listed as climate-unadjusted hazards in Table 2.2, we have not analyzed modified data sets for climate change scenarios.

Categorizing Hazard Exposure by Intensity and Likelihood

To conduct analysis of infrastructure exposure across multiple natural hazards, it is necessary to bin hazard exposure by intensity and likelihood so that hazard exposures of similar likelihood and impact can be compared with one another. To conduct this analysis, we categorized hazard intensity into two bins (lower and higher intensity) and likelihood into three categories (return periods that are less than 100 years, between 100 and 1,000 years, and greater than 1,000 years). The categorization of likelihood is cumulative because if a hazard exists at a specified intensity (e.g., two feet of flood depth) for a ten-year return period, the intensity at the 1,000-year return period will be equal to or greater than the intensity at the ten-year return period. The categorization used in this analysis to bin hazards by intensity and likelihood is presented in Table 2.4.

Table 2.4
Criteria for Binning Natural Hazards by Intensity and Likelihood

Hazard Type	Low Severity p≤100 yr	Low Severity 100 yr<p≤1,000 yr	Low Severity p>1,000 yr	High Severity p≤100 yr	High Severity 100 yr<p≤1,000 yr	High Severity p>1,000 yr
Earthquake		[0.1g,0.5g] and [0.5g,∞]; 500-yr return period	[0.1g,0.5g] and [0.5g,∞]; 500- and 2,500-yr return period		[0.5g,∞]; 500-yr return period	[0.5g,∞]; 500- and 2,500-yr return period
Landslide	All assumed low severity; all assumed p<100 yr	All assumed low severity; all assumed p<100 yr	All assumed low severity; all assumed p<100 yr			
Meteorological drought	Consider only Q95 KBDI [400,600] and KBDI [600,800]; all assumed to be p<100 yr	Consider only Q95 KBDI [400,600] & KBDI [600,800]; all assumed to be p<100 yr	Consider only Q95 KBDI [400,600] and KBDI [600,800]; all assumed to be p<100 yr	Consider only Q95 KBDI [600,800]; all assumed to be p<100 yr return period	Consider only Q95 KBDI [600,800]; all assumed to be p<100 yr	Consider only Q95 KBDI [600,800]; all assumed to be p<100 yr
Wildfire		High risk and very high risk of wildfire (fire index of 401 or higher); all assumed to be 100 yr≤p<1,000 yr	High risk and very high risk of wildfire (fire index of 401 or higher); all assumed to be 100 yr≤p<1,000 yr		Very high risk of wildfire (fire index of 1935 or higher); all assumed to be 100 yr≤p<1,000 yr	Very high risk of wildfire (fire index of 1935 or higher); all assumed to be 100 yr≤p<1,000 yr
Extreme temperature	120 degrees or higher; return periods:20	120 degrees or higher; all assumed p<100 yr	120 degrees or higher; all assumed p<100 yr	130 degrees; all assumed p<100 yr	130 degrees; all assumed p<100 yr	130 degrees; all assumed p<100 yr
Hurricane	Category 2 wind cutoff (96 mph) and Category 4 (130 mph); use natural return period	Category 2 wind cutoff (96 mph) and Category 4 (130 mph); use natural return period	Category 2 wind cutoff (96 mph) and Category 4 (130 mph); use natural return period	Category 4 (130 mph); use natural return period	Category 4 (130 mph); use natural return period	Category 4 (130 mph); use natural return period

Table 2.4—Continued

Hazard Type	Low Severity $p \le 100$ yr	Low Severity 100 yr$<p\le1{,}000$ yr	Low Severity $p>1{,}000$ yr	High Severity $p \le 100$ yr	High Severity 100 yr$<p\le1{,}000$ yr	High Severity $p>1{,}000$ yr
Riverine flooding				All assumed to be high severity; all assumed $p<100$ yr	All assumed to be high severity; all assumed $p<100$ yr	All assumed to be high severity; all assumed $p<100$ yr
Tsunami					All assumed to be high severity; all assumed 500-yr return period	All assumed to be high severity; all assumed 500-yr return period
Ice storm	Category 4 or higher; all assumed 50-yr return period	Category 4 or higher; all assumed 50-yr return period	Category 4 or higher; all assumed 50-yr return period	Category 5 for severe; all assumed 50-yr return period	Category 5 for severe; all assumed 50-yr return period	Category 5 for severe; all assumed 50-yr return period
Permanent inundation	1-ft flooding; use natural return period	1-ft flooding; use natural return period		6-ft flooding; use natural return period	6-ft flooding; use natural return period	
Tidal flooding	1-ft flooding; only 20-yr return period	1-ft flooding; only 20-yr return period	1-ft flooding; only 20-yr return period	6-ft flooding; only 20-yr return period	6-ft flooding; only 20-yr return period	6-ft flooding; only 20-yr return period
Storm surge flooding	1-ft flooding; only 100-yr return period	1-ft flooding; only 100-yr return period	1-ft flooding; only 100-yr return period	6-ft flooding; only 100-yr return period	6-ft flooding; only 100-yr return period	6-ft flooding; only 100-yr return period
Tornado			EF0 and EF3; all assumed to be $p\ge1{,}000$ yr			EF3; all assumed to be $p\ge1{,}000$ yr

CHAPTER THREE

Current Patterns of Exposure in the Continental United States

Examination of exposure data across natural hazards reveals where and how infrastructure risk may be concentrated across the country. This analysis confirms that infrastructure across the continental United States faces exposure to many types of hazards. But, the analysis also reveals areas of the country where infrastructure may face greater exposure to natural disaster, because of either exposures to multiple hazards or a disproportionate concentration of infrastructure.

Current regional patterns of exposure to natural hazards in the United States are generally believed to be well understood.[1] Figure 3.1 presents the exposure of the infrastructure assets in Table 2.1 to several individual natural hazards. In these and similar figures throughout this report, the color of the county represents the amount of infrastructure (i.e., facilities or miles of road, rail, transmission line, and pipeline) exposed to the hazards being presented. The maps in Figure 3.1 illustrate the range and variety of exposures that exist to infrastructure from natural events. Analysis of infrastructure exposure to natural hazards confirms commonly held expectations of geographic patterns for natural hazards. For example:

- Threats from ice storms, as measured by the Sperry-Piltz Ice Accumulation Index with a 50-year return interval, are most severe for the Northeast and mid-Atlantic states, although exposure extends through the Midwest.
- Exposure to riverine flooding is ubiquitous across the United States around rivers and waterways.
- Tornadoes are most likely to occur throughout the Midwest and Southeast.
- Seismic-induced tsunamis are of greatest threat in areas close to the Cascadia subduction zone in the Pacific Northwest.
- Exposure to hurricane-force winds affects the Gulf and Atlantic coasts.
- Coastal flooding is a concern across most coastal areas from extreme tidal-variations. However, Federal Emergency Management Agency (FEMA) data to support analysis of this hazard are not available for the state of Louisiana. The Gulf and Atlantic coasts are also exposed to potential flooding from storm surge.
- Exposure to seismic hazards is greatest across the Western regions, although infrequent seismic activity also exists around fault zones in the Midwest, New England, and the Southeast.
- Severe and regularly occurring meteorological drought (dryness) is prevalent in the West, South, and Southeast.

---

[1] The analysis in this report is restricted to the continental United States; for purposes of this report, the phrase "in the United States" refers to the continental United States.

Figure 3.1
Exposure of U.S. Infrastructure to Selected Individual Natural Hazards, 2015

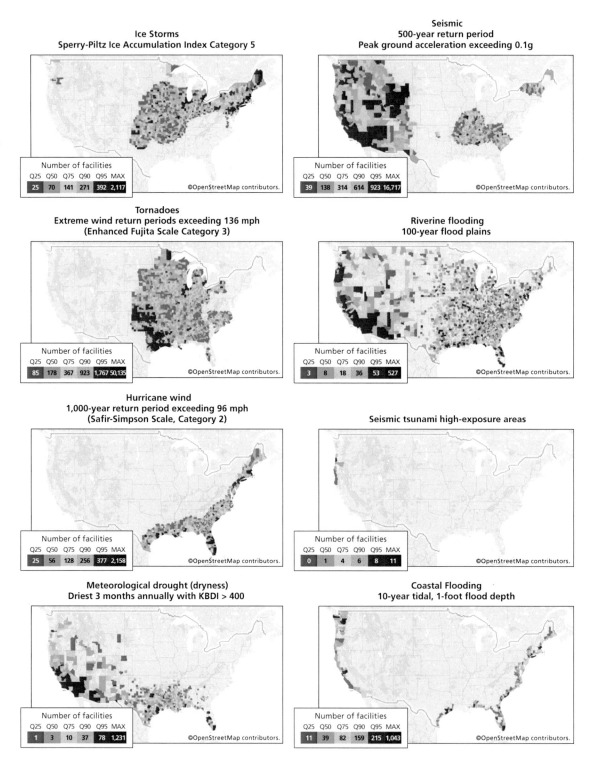

NOTE: Figure reflects gaps in hazard data, including missing data for Louisiana and other coastal areas.
RAND RR1453-3.1

Infrastructure owners and operators, economic development planners, and disaster management authorities are likely to be aware of their current exposure to individual natural hazards. However, infrastructure operations, development plans, and emergency management plans typically account for individual hazards. Single-hazard views do not reveal interactions among multiple hazards (e.g., wind damage to the electricity grid followed by a heat wave), nor would they identify the types of infrastructure that are exposed to multiple hazards. Assembling data about multiple hazards and infrastructure exposure in a consistent, geo-coded format enables identification of regions in the country where greater cumulative exposure exists to multiple natural hazards.

These data provide awareness of exposures that can inform infrastructure planning. Analyzing this cumulative exposure by intensity and likelihood (i.e., return-period) provides insight into which regions could most benefit from efforts to improve infrastructure resilience and the types of effort that would most benefit a specific region.

Most of the United States Is Exposed to Some Form of Natural Hazard

A casual scan of the maps presented in Figure 3.1 taken as a group reveals that exposure of the infrastructure assets to natural hazards of some form is common across the United States. This observation becomes even more evident when examining exposure to each type of hazard on a single map.

Figure 3.2 shows counties in which an infrastructure asset is exposed to two or more hazards, with colors used to indicate the relative amount of exposed infrastructure within a county (i.e., red indicates infrastructure with higher exposure). The analysis in Figure 3.2 presents exposure to natural hazards at the low intensity level at a return period of 1,000 years or longer using the binning approach described in Table 2.4.

Even though many counties across the United States have infrastructure exposed to more than one natural hazard, this analysis shows that not all regions face the same exposure. For example, clusters of counties with more-exposed infrastructure (i.e., shown in red) exist in several regions of the country. Larger clusters include, from west to east:

- the Pacific Northwest
- the San Francisco Bay
- Southern California
- the Mississippi River valley
- East Texas
- Chicago and its vicinities
- New York and its vicinities
- Charleston, South Carolina
- South Florida.

Figure 3.2
U.S. Infrastructure Exposed to Two or More Hazards of Lower Intensity,
Return Periods Greater Than 1,000 years

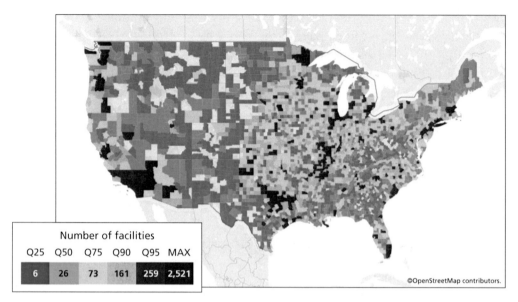

NOTE: Figure reflects gaps in hazard data including missing data for Louisiana and other coastal areas.
RAND *RR1453-3.2*

Some Regions Are Exposed to More Intense or Greater Numbers of Natural Hazards

A casual review of Figure 3.2 reveals that exposure to natural hazards is not equal across regions in the continental United States. When analysis is focused on exposure to hazards in the higher-intensity bin instead of exposure in the lower intensity bin (e.g., the analysis shown in Figure 3.2) it becomes evident that a subset of regions in the United States contain infrastructure that is exposed to two or more hazards of higher intensity. As shown in Figure 3.3, clusters of infrastructure exposed to two or more natural hazards at the higher-intensity level with a return period of 1,000 years or more exist in:

- Central and Southern California
- regions of the upper Mississippi River
- the New Madrid fault zone, which stretches to the Southwest from New Madrid, Missouri, throughout the Southern and Midwestern United States
- the Midwest in Oklahoma and Nebraska
- the mid-Atlantic coast.

Refining the analysis to focus on areas that are exposed to more than two hazards would be expected to identify a geographically smaller area. For example, Figure 3.4 shows counties with infrastructure exposed to three or more natural hazards at the higher-intensity level and at a return period of greater than 1,000 years or more. Six regions stand out from this perspective:

Figure 3.3
U.S. Infrastructure Exposed to Two or More Hazards of Higher Intensity,
Return Periods Greater Than 1,000 years

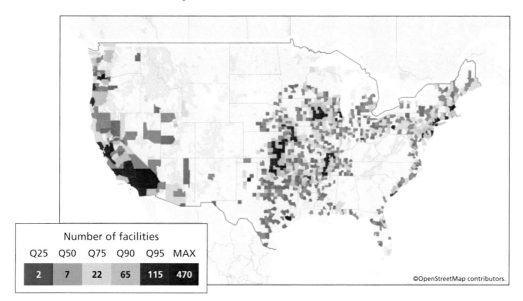

NOTE: Figure reflects gaps in hazard data including missing data for Louisiana and other coastal areas.
RAND *RR1453-3.3*

Figure 3.4
U.S. Infrastructure Exposed to Three or More Hazards of Higher Intensity,
Return Periods Greater Than 1,000 Years

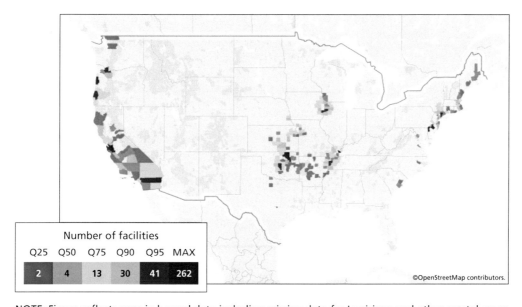

NOTE: Figure reflects gaps in hazard data including missing data for Louisiana and other coastal areas.
RAND *RR1453-3.4*

- California
- the Pacific Northwest
- regions of the upper Mississippi River
- the New Madrid fault zone
- Oklahoma
- the mid-Atlantic coast.

However, it should be noted that each of these regions has different exposure profiles. As shown in Table 3.1, the predominant types of infrastructure and the hazard exposure vary significantly across regions. The largest of the regions with the greatest variety of infrastructure is California. In addition, some of the hazards facing California (e.g., earthquakes and wildfires) are, comparatively speaking, natural hazards that have the potential for the most devastating effects.

Table 3.1
Characteristics of Infrastructure Exposure in Select Regions

Region	Predominant Infrastructure	Hazards Present	Potential for Climate Effects
California	All infrastructure sectors	Seismic, tidal flooding, riverine flooding, meteorological drought (dryness) and wildfire	Coastal flooding, drought, wildfire, and extreme temperature
Pacific Northwest	Electric power and transmission; river, interstate, and rail transportation; chemical; water	Seismic, tsunami, riverine flooding, ice storms, and meteorological drought (dryness)	Meteorological drought (dryness), wildfire and coastal flooding
Upper Mississippi River	Water, energy, transport, chemical, and nuclear	Riverine flooding, tornadoes, ice storms and meteorological drought (dryness)	Meteorological drought (dryness); exposed region extends into Illinois and Mississippi River
New Madrid Fault Zone	Rail, river, and interstate transport; power generation and transmission, gas and oil pipelines, and chemical	Seismic, ice storms, tornadoes, landslides, riverine flooding, meteorological drought (dryness), wildfires	Meteorological drought (dryness) and wildfire
Oklahoma	Interstates, rail, energy, chemical, and water	Ice storms, tornadoes, seismic, extreme temperature, riverine flooding, meteorological drought (dryness), and wildfire	Meteorological drought (dryness), wildfire and extreme temperature
Mid-Atlantic coast	Transport, electric power generation and transmission, nuclear power, pipelines, refineries, chemicals, dams, and water	Ice storms, hurricane winds, riverine flooding, tidal flooding, and storm surge	Coastal flooding

Table 3.1 also identifies which regions are exposed to natural hazards that could potentially be impacted by climate change. While exposure to natural hazards—and, in particular, exposure that may be impacted by climate change—is often conventionally assumed to be a coastal phenomenon, three of the regions exposed to three or more natural hazards of high intensity are in the Midwestern United States. In these regions, changes in precipitation patterns could exacerbate drought. Where rivers are present, changes in precipitation could also affect infrastructure that is dependent on rivers for process or cooling water, that is dependent on inland waterways for transportation, or that is located in a flood plain.

Relative Exposure to Natural Hazards

As will be discussed in Chapter Four, climate change could also potentially alter the number, type, or intensity of hazards that each of these regions face. For example, in the West and Midwest, exposure to meteorological drought (dryness),[2] wildfire, and extreme temperature may increase over the next century. Along the mid-Atlantic coast, rising sea levels may increase exposure to coastal flooding.

The relative importance of exposures across these areas depends on many factors that are not addressed by this analysis because of data limitations. Among these factors, two are worth specific mention. First, different infrastructure facilities may have different vulnerabilities to the hazards to which they are exposed depending on specific infrastructure designs, existing mitigation efforts, operational approaches, and maintenance schedules. Second, the consequences of disruptions to infrastructure may vary based on the service the infrastructure provides and the community the infrastructure serves.

While available nationwide data used in the analysis in this report do not allow for this type of in-depth analysis of infrastructure and hazard-specific vulnerability and consequence, the relative amount of infrastructure exposure and population exposure provides insights into the distribution of infrastructure exposure across the United States. To analyze this difference, we assessed the exposure of population to our set of natural hazards. To analyze population exposure, we considered the maximum exposure within a census tract to be the hazard level to the population from the 2010 U.S. census.

As an illustration, Figure 3.5 presents counties in which there is infrastructure exposed to two or more hazards of higher intensity at a return period of 1,000 years or longer. The color legend in this figure describes the difference between the relative amount of infrastructure exposure and population exposure nationwide. Counties colored orange have a greater concentration of infrastructure exposure relative to population exposure than other counties across the country. Counties colored blue have a greater concentration of population exposure relative to infrastructure exposure than other counties across the country.

Examination of the differences in concentration of population and infrastructure exposure reveals that four of the five regions described in Table 3.1 appear to have a disproportionate exposure to infrastructure (i.e., all regions except the mid-Atlantic coast). These regions with greater relative infrastructure exposure may warrant greater policy and mitigation attention.

---

[2]   *Meteorological drought* refers to conditions when dry weather patterns dominate an area. This type of drought is in contrast to hydrological drought (i.e., when low water supply becomes evident), agricultural drought (i.e., when crops are affected), or socioeconomic drought (i.e., when the supply and demand of commodities is affected).

Figure 3.5
U.S. Infrastructure Exposure Versus Population Exposure (2015) for Counties with Infrastructure Exposed to Two or More Hazards of Higher Intensity,
Return Periods Greater Than 1,000 years

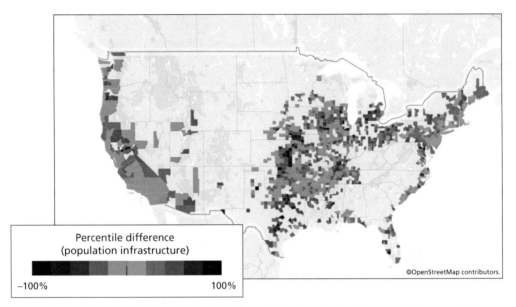

NOTE: Figure reflects gaps in hazard data including missing data for Louisiana and other coastal areas.
RAND RR1453-3.5

To mitigate events with consequences that scale with the number of people affected, such as floods, emphasis should be placed on areas like New England and the Ohio River Valley that have more-exposed populations. For mitigation that is more effective where infrastructure is denser and events do not necessarily scale with surrounding population, such as long-distance power lines or fuel pipelines, areas with more infrastructure may be more attractive. Mitigation activities for power and fuel transmissions may be important to consider in the Midwest.

CHAPTER FOUR

Climate Change and Natural Hazard Exposure

The analysis of current infrastructure exposures to natural hazard reveals patterns of exposure salient for planning today. However, because infrastructure generally is used for many decades, it is important to understand how exposures may change in the future. This requires analyzing how the effects of climate change may affect which hazards pose threats to infrastructure and how the extent of exposures may change geographically. These changes in exposure could in turn create new regions where natural disasters are concentrated and may alter the amount of infrastructure threatened.

Sources of Climate Change Data

In this analysis, we relied on natural hazard data available from the National Climate Assessment and the Intergovernmental Panel on Climate Change (IPCC). Based on their models, the research identified mechanisms that could change exposure across the United States to natural hazards. As detailed in Narayanan et al. (2016), the analysis only considers effects of climate change on natural hazards when the required data and scientific basis to do so exist; in the absence of such data and methods, hazards are treated as unchanging with climate change. The principal mechanisms for change, which are documented by Narayanan et al. (2016), are the effects of:

- sea level rise on permanent inundation, tidal flooding and coastal storm surge flooding, as measured by feet of expected flooding
- increased temperature and variations in precipitation on meteorological drought (dryness), as measured by changes in the KBDI
- the effects of increased drought on wildfire exposure, as measured by the Wild Fire Potential Index
- increased exposure to extreme daily high temperatures.

Two Forms of Uncertainty

The projected extent of changes to natural hazard exposure as a result of climate change from the above mechanisms depends on two forms of uncertainty: the extent to which greenhouse gas levels will increase in the atmosphere, and the extent to which increasing greenhouse gas levels in the atmosphere will change the climate.

The first form of uncertainty (over the future levels of atmospheric greenhouse gas concentrations) is depicted using two scenarios developed by the IPCC, referred to as Representative Concentration Pathways (RCPs). RCP 4.5 is viewed as a relatively optimistic scenario in which the global community is able to control carbon dioxide emissions. RCP 8.5 reflects the IPCC's assessment of greater atmospheric concentrations of greenhouse gases if global agreements do not achieve reduced greenhouse gas emissions and atmospheric concentrations continue to rise.

The second form of uncertainty (about the extent to which increasing greenhouse gas levels will alter the global climate and subsequently impact natural hazard exposure) is represented by variation across the climate models that are used as part of the IPCC's analysis. To illustrate this range of potential impacts, we analyze the median and most extreme (i.e., worst-case) results from these models.

Analysis of meteorological drought[1] (dryness) exposure across the country provides an illustration of how climate change is projected to change exposure to natural hazards across the continental United States, and the uncertainty in these projections. Figure 4.1 shows counties in which infrastructure is exposed to at least dry conditions (defined as KBDI values greater than 400) at some point during the driest three months of a year.

As shown in Figure 4.1, meteorological drought exposure projections for 2015 reflect conventional expectations regarding the regions of the country at greatest risk for drought, with drought exposure affecting counties primarily in the West, Southwest, South, and Southeast. These projections reflect only one type of exposure that infrastructure might face as a result of drought. For example, exposure to hydrological drought could result in shortages of water that may affect infrastructure reliant on surface and groundwater sources.

Figure 4.1
U.S. Infrastructure Expected to Be Exposed to at Least Dry Conditions at Some Point During Driest Three Months of a Year (2015 projections of KBDI Values Greater Than 400)

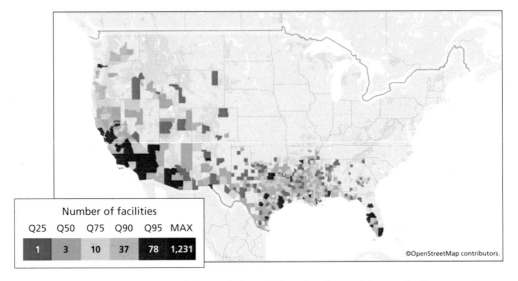

NOTE: Figure reflects gaps in hazard data including missing data for Louisiana and other coastal areas.
RAND RR1453-4.1

---

[1]   *Meteorological drought* refers to conditions when dry weather patterns dominate an area. This type of drought is in contrast to hydrological drought (i.e., when low water supply becomes evident), agricultural drought (i.e., when crops are affected), or socioeconomic drought (i.e., when the supply and demand of commodities are affected).

In the future, climate change projections suggest that exposure to meteorological drought may increase. Figure 4.2 shows the effect of both forms of uncertainty in climate change projections at different times throughout the century on projections of infrastructure that would be exposed to at least dry conditions (i.e., a KBDI value greater than 400) at some point during the driest three months of a year.

Figure 4.2 shows that in the next 25 years (i.e., present–2040), exposure to meteorological drought is projected to change dramatically only in the scenarios reflecting the greatest climate impacts and highest emissions. Projections using RCP 4.5 and RCP 8.5 with the median model for greenhouse gas impacts on the climate are similar to the projections for 2015 shown in Figure 4.1. However, meteorological drought exposure does appear to be greater using RCP 8.5 and the worst-case model. In this scenario, projections suggest exposure would expand north into other areas of the Midwest and Southeast.

As the time horizon for analysis extends further into the future, the basis for uncertainty about whether exposure will grow shifts from uncertainty over atmospheric carbon dioxide concentrations to uncertainty in model assessments. As shown in Figure 4.2, in 2065, analysis results for the RCP 8.5 scenario and median model are similar to those observed in 2040 for the worst-case model. However, in 2100, even the median model for the RCP 4.5 scenario shows expanded exposure to drought. Higher-emissions scenarios (i.e., RCP 8.5) show that infrastructure across most of the United States is projected to be exposed to dry conditions during some point in a year. Analysis of other types of drought (i.e., hydrological, agricultural, or socioeconomic drought) may reveal different patterns of exposure to infrastructure today and in the future.

Using climate change–adjusted projections also suggests that the number and size of regions of the continental United States exposed to multiple natural hazards may expand. As shown in Figure 4.3, in 2015, infrastructure in large areas along the East and West coasts and in the Midwest are exposed to two or more hazards of higher intensity at return periods of 1,000 years or longer. In contrast, the areas where infrastructure is exposed to three or more natural hazards in 2015 are significantly smaller, and almost no infrastructure is currently exposed to four or more hazards of this intensity.

However, projections for 2040 indicate that infrastructure exposure to multiple natural hazards of higher intensity could increase significantly if atmospheric concentrations of carbon dioxide continue to rise. Using the median model and RCP 8.5 scenario, projected changes in drought, extreme heat, and sea level rise (which impacts coastal flooding) expand the areas in the continental United States affected by multiple natural hazards along the coasts and throughout the Midwest and Southeast. A similar pattern is found for projections in 2065 and 2100 (not shown), and when using the RCP 4.5 emissions scenarios (not shown). Together, these projections illustrate how climate change could increase infrastructure exposure to natural hazards across large regions of the United States and could acutely affect regions along the West coast, Northeast coast, and Midwest. However, it also should be noted that these results may be lower bounds on the increase of infrastructure exposure to natural hazards because this analysis excludes some effects of climate change on natural hazards that could not be projected (e.g., the effects of changes in precipitation on riverine flooding).

Analyzing infrastructure by type at a more granular level in regard to exposure to specific natural hazards can provide a great deal of additional insight into the magnitude of current exposure by sector and how that exposure varies across hazards and the extent to which exposure in a particular sector may change as a result of climate change. Table 4.1 provides

Figure 4.2
Projections of Infrastructure Exposure to at Least Dry Conditions (KBDI values Greater Than 400) at Some Point During the Driest Three Months of a Year

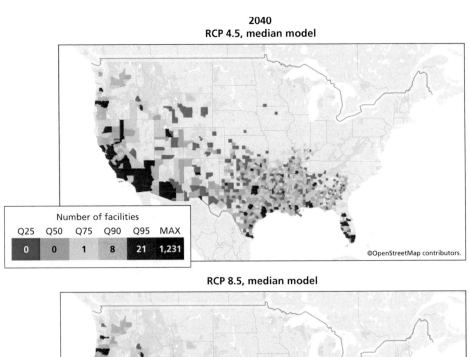

2040
RCP 4.5, median model

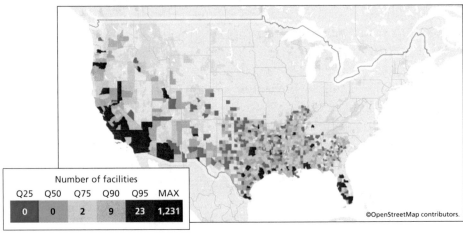

RCP 8.5, median model

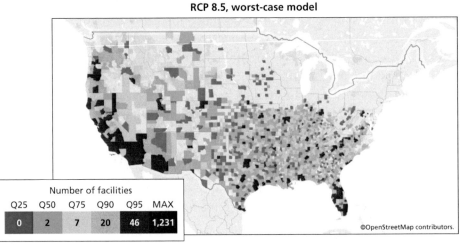

RCP 8.5, worst-case model

Figure 4.3
Projections of Infrastructure Exposure to Multiple Hazards of Higher Intensity, Return Periods Greater Than or Equal to 1,000 Years

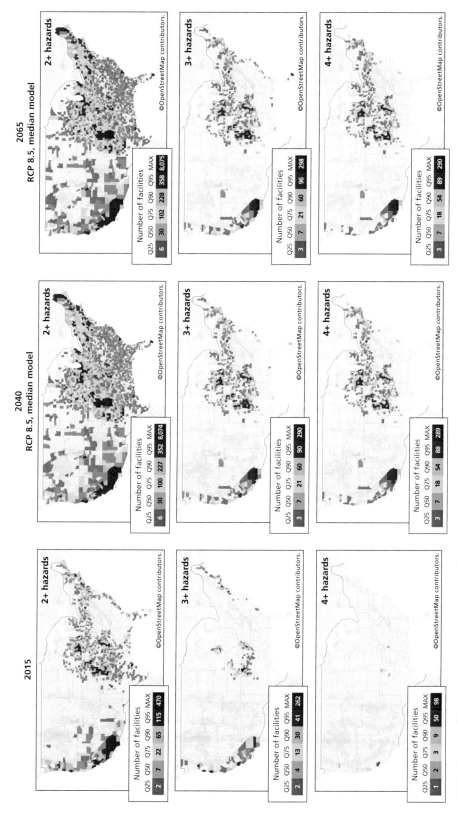

NOTE: Figure reflects gaps in hazard data including missing data for Louisiana and other coastal areas.
RAND RR1453-4.3

one example, the total miles of high-voltage and ultra-high-voltage transmission lines exposed to permanent three specific natural hazards: inundation, tidal flooding, and increased risk of wildfires. Analysis of these data relative to 2015 hazard data suggests that most of the nation's transmission infrastructure has been built in a manner such that it is not exposed to flooding and wildfires.

For example, in 2015, out of more than 200,000 miles nationwide, fewer than 15,000 miles of high- and ultra-high-voltage transmission lines are exposed to increased wildfire risk. However, as the table shows, transmission line exposure is projected to increase significantly for some hazards and scenarios in the future. For example, the amount of high- and ultra-high-voltage transmission line exposed to increased wildfire risk is projected to increase by 35 percent in 2040 using even the more moderate RCP 4.5 scenario and median climate model results. Under the RCP 8.5 scenario and worst-case model, exposure would increase 142 percent by 2100.

It also should be noted that this, and all other analysis in this report, assumes that 2015 infrastructure types and locations remain constant in the future. Exposure could increase by an even greater amount, for example, if the total miles of transmission lines increase over time.

Similar analysis can be performed regarding the number of facilities by sector that may be exposed to natural hazards in the future. Table 4.2 shows the number of water, chemical, transportation (excluding ports), energy, and communications facilities exposed today and in the future to inundation, tidal flooding, meteorological drought, and wildfire. While current exposure to these hazards in terms of number of facilities is relatively low in these sectors, a projected increase in the number of facilities that may be exposed in the future is in some cases proportionally large. For example, in the case of wildfire exposure, projected increases in the number of facilities in these sectors exposed to wildfire risk exceed 27 percent for the most optimistic scenarios analyzed (e.g., RCP 4.5, median model). However, projected increases to exposure are even larger over longer time periods and under higher emissions and higher climate change scenarios. For example, by 2100, projected increases in exposure for each of the hazards included in this analysis are close to or, in several cases, significantly exceed 100 percent in scenarios applying the RCP 8.5 emissions scenario and worst-case climate model. Collectively, these and the prior findings in this report suggest that increases in infrastructure exposure to natural hazards as a result of climate change become more certain and more extreme by 2100, but may still become an issue in some communities by 2040.

The data in Tables 4.1 and 4.2 also illustrate the uncertainty that can be introduced when interpreting climate data at low spatial resolutions. For example, the amount of infrastructure affected by wildfires appears to decline in 2040 when comparing the RCP 4.5 and RCP 8.5 emissions scenarios. Similarly, the total amount of facilities affected by wildfire appears to decrease between 2100 and 2065 for the RCP 4.5 emissions scenario. Inspection of the climate data suggests that this is the result of model results being similar, but having climate effects distributed spatially slightly differently. When combined with precise infrastructure location data, counts of exposure reflect both changes in exposure and uncertainty in the hazard data.

Table 4.1
High- and Ultra-High-Voltage Transmission Lines: Projected Miles Exposed
from Inundation, Tidal Flooding, and Wildfires, 2015–2100

Hazard/Scenario	2015	2040	2065	2100
Permanent inundation	0			
RCP 4.5, median model				600
RCP 8.5, median model			600	795 (32%)
RCP 8.5, worst-case model		600	966 (61%)	1,739 (190%)
Tidal flooding (10-year)	3,151			
RCP 4.5, median model		3,316 (5%)	3,430 (9%)	3,678 (17%)
RCP 8.5, median model		3,362 (7%)	3,574 (13%)	4,056 (29%)
RCP 8.5, worst-case model		3,614 (15%)	4,184 (33%)	5,558 (76%)
Wildfire (index value >401, driest 3 months)	14,771			
RCP 4.5, median model		19,929 (35%)	23,311 (58%)	22,384 (52%)
RCP 8.5, median model		19,280 (31%)	24,413 (65%)	32,177 (118%)
RCP 8.5, worst-case model		23,721 (61%)	28,668 (94%)	35,807 (142%)

NOTES: Figures in parentheses = percentage increases; percentage change for permanent inundation is relative to 2040 or 2065 as appropriate.

Table 4.2
Number of U.S. Water, Chemical, Transportation,[a] Energy, and Communications Facilities: Projected Exposure to Permanent Inundation, Tidal Flooding, Meteorological Drought and Wildfires, 2015–2100

Hazard/Scenario	2015	2040	2065	2100
Permanent inundation[b]	0			
RCP 4.5, median model				91
RCP 8.5, median model			91	179 (96%)
RCP 8.5, worst-case model		91	283 (211%)	1,421 (1,462%)
Tidal flooding (10-year)	3,413			
RCP 4.5, median model		3,585 (5%)	3,728 (9%)	4,037 (18%)
RCP 8.5, median model		3,643 (7%)	3,865 (13%)	4,441 (30%)
RCP 8.5, worst-case model		3,942 (15%)	4,563 (34%)	6,012 (76%)
Drought (75th percentile, index value >401)	15,710			
RCP 4.5, median model		19,089 (22%)	19,942 (27%)	21,100 (34%)
RCP 8.5, median model		19,833 (26%)	26,553 (69%)	34,121 (117%)
RCP 8.5, worst-case model		31,373 (100%)	45,032 (187%)	53,440 (240%)
Wildfire (index value >401)	4,991			
RCP 4.5, median model		6,328 (27%)	7,550 (51%)	6,999 (40%)
RCP 8.5, median model		5,946 (19%)	8,290 (66%)	10,743 (115%)
RCP 8.5, worst-case model		7,819 (61%)	9,238 (85%)	13,121 (163%)

[a] Ports excluded.
[b] Percentage change for permanent inundation is relative to 2040 or 2065 as appropriate.

CHAPTER FIVE

Findings and Policy Considerations

Unlike many prior assessments that focused on individual infrastructure sectors or individual hazards,[1] the analysis in this report and the accompanying documentation on underlying data and methods (Narayanan et al., 2016) provide an integrated view of infrastructure exposure to a range of potentially high-intensity natural hazards. The analysis in this report describes exposures across multiple infrastructure sectors and multiple hazards and incorporates both current infrastructure exposure to natural hazards and uncertainty about the extent and intensity of natural hazards in the future in the context of projected changes to climate patterns.

Key Findings

Several key findings emerge from this analysis:

Infrastructure exposure to natural hazards is expected to increase—and, in some cases, increase substantially—across the continental United States. Almost all infrastructure assets in the continental United States are currently exposed to some form of natural hazard. Specific exposure is largely dependent on the current geographic patterns of natural hazard phenomena. However, many of these patterns are not projected to be static. Even the most optimistic projections of changes to sea level rise, precipitation, and extreme temperatures suggest that more infrastructure assets—in more places throughout the country—will be exposed to more natural hazards of high intensity.

Infrastructure in some areas of the country currently faces disproportionate exposure to natural hazards, and this exposure is likely to increase in the future as a result of climate change. Both infrastructure and natural hazard exposure have clustered geographic distributions, which are found to overlap in a number of key regions of the continental United States when infrastructure exposure is analyzed across multiple infrastructure sectors and natural hazard types. Examples of regions where such disproportionate exposure exists include California, the Pacific Northwest, the upper Mississippi River, the New Madrid fault zone, regions in Oklahoma, and the mid-Atlantic coast (see Table 3.2). Notably, three of these areas are in the Midwest despite broader awareness of exposures to coastal regions from natural hazards. Regions with potentially greater current and/or future exposure to natural hazards can only be identified by evaluating multiple types of natural hazard exposures together and incorporating the best available information regarding the uncertainty of climate change impacts on

---

[1]   Examples include analysis of the resilience of energy distribution infrastructure in the Quadrennial Energy Review (DOE, 2015) on a single sector or flood risk mitigation studies conducted for FEMA on climate change impacts on flood areas (AECOM, 2013).

natural hazard exposures in the future. These findings are likely to suggest several implications for policy and future research. In the short term, several are apparent:

Infrastructure and community resilience efforts should incorporate potential impacts from climate change and potential increases in exposure to natural hazards. Any improvements in infrastructure resilience should take into account changes in future natural hazard exposure, including but not limited to the results of the analysis in this report. Climate change–induced natural hazard exposure should be accounted for when planning to recapitalize existing infrastructure, expand infrastructure to meet shifting or growing demands, update infrastructure based on technological advancements, replace infrastructure damaged by disasters, or make major decisions regarding infrastructure investment.

More granular information is needed about specific natural hazard exposure and infrastructure in a community to respond effectively to climate change–induced natural hazard exposure changes. Ultimately, assessing and responding to the specific hazards, infrastructure type, location, condition, and the resulting infrastructure vulnerability in that region, whether in a geographic cluster with high exposure or a region with less exposure, can improve the resilience of communities and infrastructure. However, understanding that a region may be exposed to multiple natural hazards, now and perhaps increasingly so in the future, is the first step and may help to identify mitigation policies and investments that would not have been identified or would not appear to be cost effective if a single-hazard, single-sector, or short-term planning perspective were employed.

Data Gaps

The exposures presented in this report are lower bound on current and future exposures because of limits in infrastructure data, natural hazard data, and understanding of the effects of climate change. Better and more complete data are needed to support regional infrastructure resilience planning. This analysis identified several gaps in available data and existing knowledge. Each of these gaps, which have been briefly discussed in the preceding sections of this report, represent opportunities for further research or data collection to support regional resilience planning. The most significant data gaps include:

- **Incomplete baseline data for several hazards and areas. Incomplete data exist for several key hazards.** For example, data for riverine flooding are based on digitized FEMA designations of 100-year flood plain data; similarly, coastal surge flooding is based on digitized FEMA advisory base flood elevation data. These data could be improved in several ways. First, data are only available where FEMA has certified the flood maps, but many areas do not yet have FEMA–certified flood maps, creating significant gaps. As an example, advisory base flood elevations are only available from FEMA for selected areas. Data are not available for the state of Louisiana and significant gaps exist along the Florida coastline. While some of these data have been developed for regional studies, these have not been incorporated into national data sets. As additional maps become available, the data gap will be reduced and provide an opportunity for improved analysis. Second, the 100-year flood plain designations do not specify flood depths within the designated area, and analysis has not been done to convert changes in precipitation patterns into changes in expected flood depths. While analysis has been done to convert changes in precipitation into changes in stream flow, the additional analysis to convert stream flow to flood depth would allow for greater analysis of current and future flood damage.

- **Lack of probabilistic exposure data for several hazards.** Assessments of exposure from ice storms, wildfires, landslides, and meteorological drought are all based on indexes that communicate propensity for intense hazard phenomena to occur (e.g., return periods). However, the indexes are not mechanistically associated with consequences of these events, have not been shown to be correlated to damages, and are not probabilistic. In each case, additional research could improve understanding of natural hazard exposure to support disaster mitigation planning and investment. Early efforts on improving this scientific knowledge would be best directed toward those hazards that are projected to have the greatest future impact, such as drought.

- **Uncertainty about the effects of climate change on many hazards.** Characterizing the effects of climate change on local and regional exposures to natural disasters is an active area of scientific inquiry. This analysis only reflects projections of climate change for a subset of hazards for which climate change could affect exposures. As data and models improve characterization of future hazards, the exposures described in this report may increase. This fact is especially salient for exposure to riverine flooding and hurricane winds. Both of these hazards affect large regions of the United States, are known to damage infrastructure, and are expected to be affected by climate change. Yet, current data and models do not characterize how exposures to these hazards will change.

- **Inadequate detail for infrastructure data to support risk analysis.** As noted above, the data available to support a national assessment of infrastructure exposure to natural hazards do not contain the level of detail needed regarding infrastructure assets to progress beyond simply conducting an assessment of exposure to conducting a true assessment of risk from those exposures. Specifically, three types of information are missing.

  - The HSIP Gold data used in this analysis contain no information about the condition of infrastructure (e.g., maintenance records). The data also do not contain information about mitigation countermeasures that might be in place, such as flood walls or structural reinforcement to protect against high wind or earthquakes. These data would be necessary to support detailed vulnerability analysis.

  - Vulnerability curves that describe whether a facility would be damaged or disrupted when exposed to natural hazards of different intensities do not exist for most types of infrastructure. The most complete data of this type exists in the HAZUS model (FEMA's methodology for estimating potential losses from disasters), but these data are limited to hurricane winds, earthquakes, and floods, for limited types of infrastructure. These data are also needed to support detailed vulnerability analysis.

  - The HSIP Gold data do not provide complete and valid information about the level of service provided or extent of area served by an infrastructure asset. This information would be a significant component of what is needed to understand the consequences of disruption.

While all of these data would be needed to extend an exposure analysis into a risk analysis, assembling national data sets of this information for all hazards and infrastructure would surely be a complicated and labor-intensive endeavor. An alternative approach to filling this data gap at the national level would be to develop more-detailed data at a regional or local level in areas where exposure analysis suggests further assessment of mitigation efforts is most needed.

- **Less data available for Alaska, Hawaii, and U.S. territories.** Limits on the resolution and availability of data led us to exclude analysis of infrastructure exposures outside the continental United States. This limitation is especially salient for analysis of coastal flooding, which relies on National Oceanic and Atmospheric Administration sea level gauges. To support infrastructure and resilience planning nationwide, the infrastructure and hazard data sets used in this analysis must be extended to these states.

- **Uncertainty about where and how future infrastructure development will occur. This analysis incorporates a static picture of infrastructure.** It does not reflect projections on how future infrastructure will be developed to meet the U.S. demands as populations increase and shift or as new technologies emerge. Future analysis could consider the implications of infrastructure exposure to natural hazards for alternative future infrastructure development scenarios.

Each of these observations provides a basis for planning as the federal government, state and local governments, and the private sector seek to improve resilience against current and future exposure to high-intensity natural hazard. Addressing these issues effectively and comprehensively will require a multifaceted effort that must include collecting descriptive information from and about communities, improving scientific knowledge about hazard phenomena, and developing tools and institutions to plan mitigation strategies for the complex and uncertain array of natural hazards that do and may increasingly threaten communities across the nation.

Interactions Between Infrastructure and Hazards

The following table indicates the physical interactions reflected in the exposure analyses presented in this report. Additional details about how these assessments were made are provided in the technical documentation supporting the natural hazard and infrastructure data sets (Narayanan et al., 2016).

Table A.1
Interactions Leading to Physical Infrastructure Damage Between Infrastructures and Hazards

Infrastructure Name	Winter Storm	Extreme Temp.	Hurricane	Landslide	Riverine	Storm Surge	Wildfire	Digital Coast SLR	Tidal Flood	Tsunami	Meteorological Drought	Quake	Tornado
Interstates	0	0	0	1	0	0	0	0	0	1	0	1	0
Road bridges	0	0	1	1	1	1	1	0	0	1	0	1	1
Road tunnels	0	0	0	1	0	0	0	0	0	1	0	1	1
Fixed-guideway transit systems transit lines	0	0	1	1	1	0	1	0	0	1	0	1	1
Electric power generation plants	0	1	1	1	1	1	1	1	1	1	1	1	1
Fixed-guideway transit systems stations	0	0	1	1	1	1	1	1	1	1	0	1	1
Railroad transit lines	0	0	0	1	1	0	0	0	0	1	0	1	1
Railroad stations	0	0	1	1	1	1	1	1	1	1	0	1	1
Railroad bridges	0	0	1	1	1	1	1	0	0	1	0	1	1
Railroad tunnels	0	0	0	1	0	0	0	0	0	1	0	1	1
Railroad yards	0	0	0	1	1	1	1	1	1	1	0	1	0

Table A.1—Continued

Infrastructure Name	Winter Storm	Extreme Temp.	Hurricane	Landslide	Riverine	Storm Surge	Wildfire	Digital Coast SLR	Tidal Flood	Tsunami	Meteorological Drought	Quake	Tornado
Airports	0	0	1	1	1	1	1	1	1	1	0	1	1
FAA air route traffic control centers	0	0	1	1	1	1	1	0	0	1	0	1	1
Intermodal terminal facilities	0	0	1	1	1	1	1	1	1	1	1	1	1
U.S. coastal, great lakes and inland ports	0	0	1	1	1	1	0	1	1	1	0	1	1
Canals	0	0	0	1	0	0	0	0	0	1	0	1	1
Locks	0	0	0	1	0	0	0	0	0	1	0	1	1
Internet exchange points	0	0	1	1	1	1	1	1	1	1	0	1	1
Chemical industries	0	1	1	1	1	1	1	1	1	1	1	1	1
Oil and natural gas pipelines	0	0	0	1	0	0	0	0	0	1	0	1	1
Petroleum, oil, and lubricants storage facilities	0	0	1	1	1	1	1	1	1	1	0	1	1
Refineries	0	1	1	1	1	1	1	1	1	1	1	1	1

Table A.1—Continued

Infrastructure Name	Winter Storm	Extreme Temp.	Hurricane	Landslide	Riverine	Storm Surge	Wildfire	Digital Coast SLR	Tidal Flood	Tsunami	Meteorological Drought	Quake	Tornado
Natural gas import/export points	0	0	1	1	1	1	1	0	0	1	0	1	1
Natural gas processing plants	0	0	1	1	1	1	1	0	0	1	0	1	1
Energy distribution and control facilities	0	1	1	1	1	1	1	1	1	1	0	1	1
Nuclear plants	0	1	1	1	1	1	1	1	1	1	1	1	1
Nuclear fuel facilities	0	1	1	1	1	1	1	1	1	1	0	1	1
Substation	0	1	1	1	1	1	1	1	1	1	0	1	1
Electric power transmission lines	1	0	1	1	1	1	1	0	0	1	0	1	1
Dams	0	0	0	1	0	0	0	0	0	1	0	1	1
Population	1	1	1	1	1	1	1	1	1	1	1	1	1
Wastewater treatment plants	0	1	0	1	1	1	1	1	1	1	1	1	1

NOTE: A cell denoting "1" indicates that the infrastructure is affected by the specified hazard. On the contrary, "0" indicates that that infrastructure is not affected by the noted hazard.

Abbreviations

DHS	U.S. Department of Homeland Security
DOE	U.S. Department of Energy
FEMA	U.S. Federal Emergency Management Agency
HSIP	Homeland Security Infrastructure Protection
HUD	U.S. Department of Housing and Urban Development
IPCC	Intergovernmental Panel on Climate Change
KBDI	Keetch-Byram Drought Index
RCP	Representative Concentration Pathway
SLR	sea level rise

References

AECOM, *The Impact of Climate Change and Population Growth on the National Flood Insurance Program Through 2100*, prepared for the Federal Insurance and Mitigation Administration, Federal Emergency Management Agency, June 2013.

American Society for Civil Engineers, *Report Card for America's Infrastructure, American Society for Civil Engineers*, 2013.

ASCE—*See* American Society for Civil Engineers.

DHS —*See* U.S. Department of Homeland Security.

DOE—*See* U.S. Department of Energy.

Finucane, Melissa L., Noreen Clancy, Henry H. Willis, and Debra Knopman, *The Hurricane Sandy Rebuilding Task Force's Infrastructure Resilience Guidelines: An Initial Assessment of Implementation by Federal Agencies*, Santa Monica, Calif.: RAND Corporation, RR-841-DHS, 2014. As of April 6, 2016:
http://www.rand.org/pubs/research_reports/RR841.html

Groves, David G., Martha Davis, Robert Wilkinson, and Robert Lempert, "Planning for Climate Change in the Inland Empire: Southern California," *Water Resources IMPACT*, Vol. 10, No. 4, July 2008, pp. 14–17.

Groves, David G., Jordan R. Fischbach, Evan Bloom, Debra Knopman, and Ryan Keefe, *Adapting to a Changing Colorado River: Making Future Water Deliveries More Reliable Through Robust Management Strategies*, Santa Monica, Calif.: RAND Corporation, RR-242-BOR, 2013. As of April 1, 2016:
http://www.rand.org/pubs/research_reports/RR242.html

Haimes, Y. Y., "On the Definition of Vulnerabilities in Measuring Risks to Infrastructures," *Risk Analysis*, Vol. 26, No. 2, 2006.

Homeland Infrastructure Foundation–Level Data Subcommittee Online Community, "Welcome to the HIFLD Subcommittee Home Page," web page, undated. As of April 7, 2016:
https://gii.dhs.gov/hifld

HUD—*See* U.S. Department of Housing and Urban Development.

LACPRA—*See* Louisiana Coastal Protection and Restoration Authority.

Louisiana Coastal Protection and Restoration Authority, *Louisiana's Comprehensive Master Plan for a Sustainable Coast*, 2012. As of April 8, 2016:
http://coastal.la.gov/a-common-vision/2012-coastal-master-plan/

Narayanan, Anu, Henry H. Willis, Jordan Fischbach, Drake Warren, Edmundo Molina-Perez, Chuck Stelzner, Katie Loa, Lauren Kendrick, Paul Sorenson, and Tom LaTourrette, *Characterizing National Exposures to Infrastructure from Natural Disasters—Data and Methods Documentation*, Santa Monica, Calif.: RAND Corporation, RR-1453-1-DHS, 2016.

New York City Department of Environmental Protection, *Jamaica Bay Watershed Protection Plan, 2014 Update*, October 2014. As of April 8, 2016:
http://www.nyc.gov/html/dep/pdf/jamaica_bay/jbwpp_update_10012014.pdf

NYCEP—*See* New York City Department of Environmental Protection.

U.S. Department of Energy, *Quadrennial Energy Review, First Installment*, Washington, D.C., 2015. As of August 6, 2015:
http://energy.gov/epsa/quadrennial-energy-review-qer

U.S. Department of Homeland Security, *National Infrastructure Protection Plan 2013—Partnering for Critical Infrastructure Security and Resilience*, Washington, D.C., 2013. As of August 6, 2015:
https://www.dhs.gov/national-infrastructure-protection-plan

U.S. Department of Housing and Urban Development, *Hurricane Sandy Rebuilding Strategy*, Washington, D.C., August 2013. As of August 6, 2015:
http://portal.hud.gov/hudportal/documents/huddoc?id=hsrebuildingstrategy.pdf

The White House, Presidential Policy Directive 21—Critical Infrastructure Security and Resilience, Office of the Press Secretary, February 12, 2013. As of August 6, 2015:
https://www.whitehouse.gov/the-press-office/2013/02/12/presidential-policy-directive-critical-infrastructure-security-and-resil

Willis, Henry H., and Kathleen Loa, *Measuring the Resilience of Energy Distribution Systems*, Santa Monica, Calif.: RAND Corporation, RR-883-DOE, 2015. As of April 01, 2016:
http://www.rand.org/pubs/research_reports/RR883.html